Building a Home

Please visit our website at: www.garethstevens.com
For a free color catalog describing Gareth Stevens' list of high-quality books and multimedia programs, call 1-800-542-2595 (USA) or 1-800-461-9120 (Canada). Gareth Stevens Publishing's Fax: (414) 332-3567.

Library of Congress Cataloging-in-Publication Data available upon request from publisher. Fax: (414) 336-0157 for the attention of the Publishing Records Department.

ISBN 0-8368-2814-3

This North American edition first published in 2001 by
Gareth Stevens Publishing
330 West Olive Street, Suite 100
Milwaukee, WI 53212 USA

Created and produced as *So Many Ways to Build a Shelter* by

QA INTERNATIONAL
329 rue de la Commune Ouest, 3e étage
Montréal, Québec
Canada H2Y 2E1
Tel.: (514) 499-3000 Fax: (514) 499-3010
www.qa-international.com

Printed in Canada

1 2 3 4 5 6 7 8 9 05 04 03 02 01

Gareth Stevens Publishing
A WORLD ALMANAC EDUCATION GROUP COMPANY

A home of their own

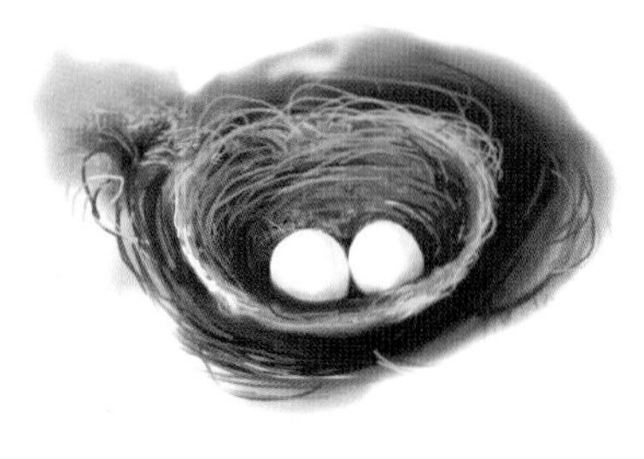

Animals use shelters for many reasons: to hide from predators, to escape bad weather, or to care for their young. Some animals take refuge at the tops of trees, in cracks in rocks, or in the undergrowth. Others build comfortable homes. They use their legs, claws, beaks, or teeth. For building materials, they use leaves, twigs, wood, earth, hair, down, wool, moss, and even pieces of plastic and fabric. Some animals make their own building materials.

Saliva nests

Swiftlets use strands of their own saliva to make their cup-shaped nests. During the mating season, the salivary glands under their tongue produce long, thin, sticky filaments. The saliva hardens and quickly sticks to the walls of the rocky caves the birds call home.

swiftlet

Are you curious?

Every year, men risk their lives to gather swiftlet nests, an Asian delicacy. In Hong Kong, a single bowl of swiftlet's nest soup can cost as much as $60!

case-bearing leaf beetle

An original house

Many small beetles have homes made entirely of excrement. As soon as the eggs are laid, the parents roll them in their body wastes until the eggs look like little dried fruits. When the eggs hatch, the young not only hang on to their coating, they enlarge it by adding their own excrement.

Living in an air bubble

The water spider lives under the water. It spins a watertight globe of threads that it attaches to water grasses. During its trips to the surface of the water, the spider traps small air bubbles in the hairs of its legs and claws. It then fills its underwater home with the air bubbles. They push out the water, and the spider is safe in its air-filled home.

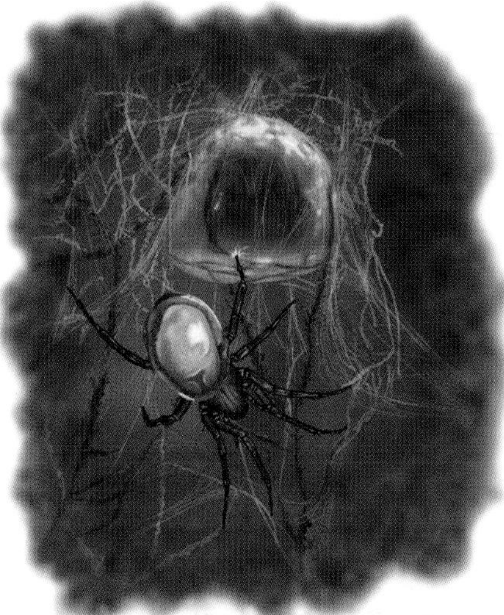

water spider

A platform over the sea

The lesser noddy is a sea bird that lives on tropical islands of the Atlantic, Pacific, and Indian Oceans. It often builds its home on a rocky ledge. The home is made from the bird's own feathers and excrement. The bird mixes these materials together and stomps on them until they form a solid platform where the family can make itself at home.

lesser noddy

Ready-made homes

Building a house takes a great deal of effort and energy. That is why some animals look for a home that is already made. Nature often provides natural shelters. A hollow tree trunk may be inhabited by owls, weasels, lemurs, or parrots. A damp cave can be a home to bats, insects, birds, and bears. Cracks in rocks can house lizards, snakes, amphibians, and insects. Some small animals prefer structures made by humans. A chimney or a crack in the wall of a house can be a perfect place for them to live. Many animals move into structures built by other animals.

An owl among woodpeckers

The tiny elf owl lives in the deserts of the southern United States. Active all night long, it is always hunting for food: grasshoppers, moths, and beetles. At daylight, it settles into a hole drilled into a giant cactus by a woodpecker. Perfectly safe, it enjoys a well-deserved rest.

elf owl

Are you curious?

The eyes of owls cannot move. To look in a different direction, these animals have to turn their whole head. Some can turn their heads three quarters of the way around!

red-beaked hornbill

Trapped in a tree

The hornbill lives in tropical forests. It uses excrement, wet soil, rotten wood, and saliva to make a sort of plaster. The female plugs up a natural hole in a tree trunk with this mixture. For two months, her mate slides small fruits and insects through a narrow slit to feed mom and the kids. The snakes and monkeys that prey on hornbills cannot see or hear the young birds.

Rock shelters

Although many species of parrots make their nests in holes in trees, the bahamian amazon nests underground. The ground on the West Indian island of Abaco, its natural habitat, is made up of limestone in which water has already carved out deep cavities. The water has receded, but the holes remain. They make perfect burrows for laying eggs and raising young amazons.

bahamian amazon

How sly is the fox?

Foxes are found on almost every continent: North America, Europe, Africa, Asia, and even Australia. Although it sometimes digs its own dwelling, this beautiful mammal often moves into the burrows of other animals. It especially likes badger burrows, which are found on sunny slopes.

red fox

Underground shelters

There are not enough natural burrows in the wild to house all the animals that need homes. Most burrowing animals have to dig their own holes. The burrow of a hippopotamus is usually no more than an ordinary hole in the ground, shaped like a comfortable bathtub. The craters dug by female boars are a little deeper and allow their young to rest in safety. Many animals bury themselves in the ground: cockles, sea urchins, sand fleas, snails, and insect larvae.

A hidden home

The trap-door spider uses its hooks to dig its tube-shaped burrow in very dry earth. The burrow is lined with silk and kept hidden by a plug attached to the ground by a thread of silk. The spider can open or close the cover of its home at any time. If an insect gets too close, the spider opens the cover and pounces on it!

trap-door spider

Are you curious?

Trap-doors are large spiders found mainly in tropical regions. Certain species attack very large prey. They have been seen capturing small birds, frogs, lizards, and even snakes.

The phantom of the beaches

The phantom crab hollows out its U- or Y-shaped home in the sand of certain beaches. It comes out at night to feed on debris and live prey left behind by the high tide. It spends the day at the bottom of its home. The camouflaged entrance is blocked with sand.

ghost crab

Atlantic puffin

Burrowing birds

The Atlantic puffins found in the cold seas of North America and Europe come ashore only to reproduce. These birds live in burrows they dig with their own beaks and claws. The thousands of couples that make up a single colony dig so many holes along the shoreline that the ground sometimes caves in!

A U-shaped hiding place

Where do those little twisted pieces of earth come from? The lugworm is a freshwater worm that lives completely camouflaged in the sand where it digs its U-shaped hiding place. The lugworm feeds on the tiny animals and plants in its sandy home. Every 40 minutes or so, it drops its body wastes in little piles at the door of its home.

lugworm

Excellent potters

Earth is a very useful natural material. It can be mixed with saliva, or even with excrement, to make dense mud. Animals that use earth build solid homes that usually last a long time. The mixture they use becomes as hard as plaster when it dries. Many birds make little balls of clay that they stick together by using their saliva as mortar. Insects also use dried mud to build homes of different shapes.

The heaviest nest

The rufous hornero gathers 1,500 to 2,500 little piles of clay to build its ball-shaped nest. It then assembles them with grass, feathers, and cow dung. These little mud houses are often found on branches or at the top of a post. They take 10 to 15 days to build, can measure as much as 8 inches (20 centimeters) in diameter, and can be used only once!

Are you curious?

The rufous hornero — a South American bird — is known as the baker bird because its round home looks like a little oven. It becomes so hot within twenty to twenty-six days that the bird must abandon it to avoid suffocating!

A pool of mud

Some tree frogs build mud shelters that look like miniature pools. The male of one species builds a circular pool about 12 inches (30 cm) in diameter, surrounded by a wall 4 inches (10 cm) high. The couple deposits its eggs in the pool, where they are safe from predators who cannot climb over the wall of mud.

tree frog

greater flamingo

A mud cone

The pink flamingos of Europe, Africa, and the Middle East live in huge colonies. To make their nests, the males and females gather mud, stones, shells, feathers, and grass. The flamingos shape, stamp, and sleep on their nests until the materials are packed down. The cone-shaped nests can reach heights of nearly 16 inches (40 cm).

potter wasp

rufous hornero

Potter wasps

The potter wasp forms small urns from little balls of clay it makes after it rains. In each of these vases, the female lays a small egg that is suspended from the ceiling by a thread of silk. A caterpillar or a paralyzed insect is placed at the bottom of the urn to provide food for the larva.

Tunnel builders

Some animals do not dig ordinary holes in the ground. They build networks of tunnels with birthing rooms, bedrooms, resting areas, and toilets. These underground homes can house as many as several hundred animals and extend over a very large area. Some animals rarely leave their underground dwellings. They sleep, store supplies, reproduce, raise their young, and feed there. They have a cool, damp place to live in all year long.

Underground towns

Prairie dogs build the largest underground towns. Each town is made up of 50 to 100 burrows. Each burrow is home to about 15 members of one family. The burrows have a vertical corridor that leads to a series of tunnels ending in various rooms. Watchdogs stand guard at the entrances to the burrows. They perch on little hills that are made from the discarded soil from the tunnels. At any sign of danger, they let out a yelp to warn the others to scoot back into their burrows.

black-tailed prairie dogs

Are you curious?

Prairie dogs are not related to actual dogs. They are actually hare-sized rodents found on the prairies of North America. They are referred to as dogs because their yelp sounds very much like a bark.

A two-story house

The home of the pocket gopher has two stories. On the first, deeper level are little rooms and storage areas. On the second are the bedroom and the long feeding corridors. This hamster-sized rodent appears happy to be a burrowing animal. It continues to dig tirelessly even after it snows.

Eastern pocket gopher

Old World rabbit

A place for everything

Cottontail rabbits live in colonies of about 100 animals. They dig branched burrows for shelter. When the female is ready to give birth, she digs a special hole known as a "nursery burrow" away from the rest of the colony. This hole, about 20 to 28 inches (51 to 71 cm) deep, becomes home to the baby rabbits.

A temporary burrow

The aardvark has pointed ears, a snout like a pig's, a tail like a kangaroo's, and a body covered with stiff hairs. As it makes its way across African savannahs, this mammal digs burrows to escape the intense heat. Small rodents, birds, and snakes move into these burrows when the aardvark moves on.

aardvark

More tunnel builders

Animals that spend most of their time underground have adapted to this darkest possible living space. They don't really need eyes, so their eyes are small or are missing entirely. Also, since the flaps of ears can be a nuisance underground, they are either very small or completely absent. To find their way around and search for food, burrowing animals rely on their well-developed senses of smell, hearing, and touch.

A champion burrower

The mole digs tunnels with its claws and forelegs, while its rear legs push the earth to the surface. Little mounds are formed by the excess earth. Mole holes have an opening on the outside to let air in and out. Beneath the biggest mole hole is the main room, an area carpeted with leaves, where moles rest after hunting.

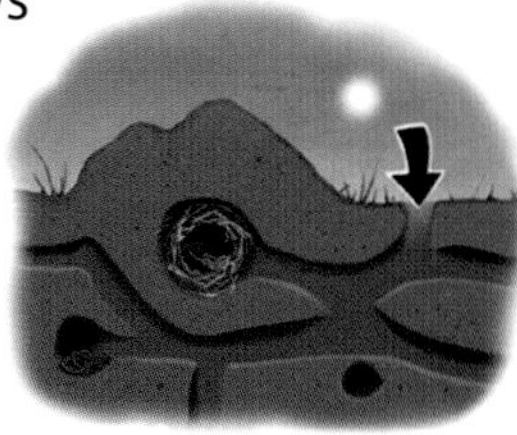

European mole

American badger

A very clean burrow

The American badger is a carnivorous mammal with short, powerful legs. It is a champion burrower. Its 30-foot- (10-meter-) long burrows are very clean. The rooms are carpeted with a layer of soft grass that the badger changes frequently.

Shelter from the sun

The wombat is a marsupial. It uses its teeth and claws to dig its burrow through clay soil that is as hard as rock. The wombat spends the entire day in its burrow, where the temperature is just right. The wombat may share its shelter with snakes, lizards, birds, and rabbits.

common wombat

naked mole rat

Life in the dark

This odd-looking blind animal with its pink, wrinkled skin, short legs, and flat head will never win any beauty contests. It rarely leaves its underground dwelling, which is made up of long tunnels the mole rat digs with powerful front teeth. The teeth remain outside of its mouth even when it is closed.

Are you curious?

Moles do some amazing things. Champion diggers, they can move 13 pounds (6 kilograms) of earth in barely twenty minutes!

Weavers

Some animals are weavers. Using their beaks, limbs, or teeth, they gather leaves, grass, and silky threads to make finely crafted little homes. Birds are excellent weavers. Interlacing blades of grass with great skill, they knot, weave, and braid to create lovely homes. They use long, flexible, and sturdy materials, such as climbing plants, roots, ribbon-shaped leaves, and even torn narrow strips of broad leaves.

Woven houses

Weaverbirds are the best weavers in the animal kingdom. Their nest starts out as a simple ring of interwoven grass. The male bird then adds and weaves in plant materials. There are about seventy species of weaverbirds in Africa and Asia. All their nests look like little purses hanging from trees.

masked weaver

Are you curious?

Among weaverbirds, nest building is a way of attracting a partner. The female inspects the work done by the male. If she likes the nest, she accepts the male as her companion.

A quick builder

It takes the harvest mouse only five to ten hours to build its pretty woven nest. Using its teeth and legs, the rodent tears grass into thin strips, which it weaves into a ball. The inside of the nest is carpeted with leaves and moss. The nest is perched about 3 feet (1 m) above the ground on a long stalk of wheat or stem of grass.

Old World harvest mouse

long-tailed tailor-bird

A big designer

The tailor-bird builds its nest from broad leaves and spider silk. It does this by folding two leaves together. Using its beak as a needle, it pierces little holes along the edges of the leaves and sews them together with the spider silk. The inside of the leaf is carpeted with a soft, warm layer of plants, sheep's wool, and animal hair.

Designing larvae

Weaver ants work in teams to build their pretty homes. While several ants hold the leaves in place, others walk young larvae across the joined leaves. The larvae produce threads of silk in special glands. The thousands of threads of silk are arranged to provide solid support for the walls of this unusual dwelling.

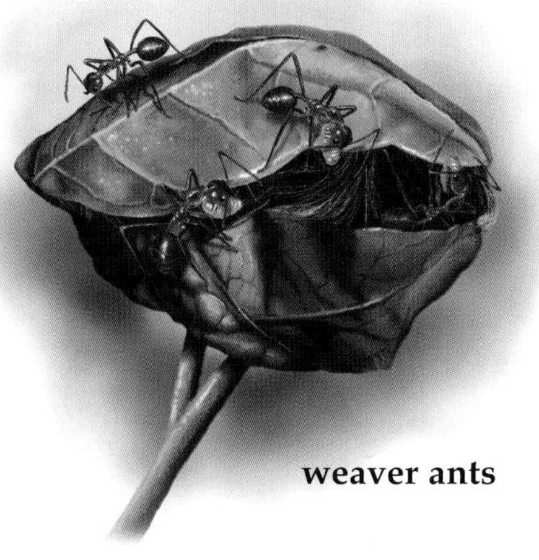

weaver ants

Master drillers

Animals that build their own homes dig holes in the ground, shape earth, or weave and knot twigs. Their work requires time, energy, strength, and skill. Just hollowing out dirt to make a burrow must be difficult and tiring. Some animals perform tasks that require great strength. They are known as borers. They carve out homes in hard substances such as wood or rock. The beaks and powerful jaws of certain small animals do work that human beings would not be able to do with their bare hands.

Tireless drillers

Bee-eaters hollow out their homes in a sandy riverbank, the slope of a hill, or a rock face. Hovering next to one of these spots, the bird makes an opening with its pointed, sturdy beak and keeps on digging for days on end. At the end of the tunnel, the bee-eater makes a small, down-covered room for its young birds.

carmine bee-eater

Are you curious?

Bee-eaters found in Europe, Africa, Asia, Australia, and the Philippines are not the only birds that live in burrows. Kingfishers and bank swallows also live in burrows.

Tunneling through wood

The female goat-moth lays hundreds of eggs in the cracked bark of a leafy tree. Fourteen days later, small caterpillars with powerful jaws emerge from the eggs. For over two years, they dig tunnels in the wood of the tree, where they will be protected from their enemies. These tunnels can block the flow of sap and kill the tree.

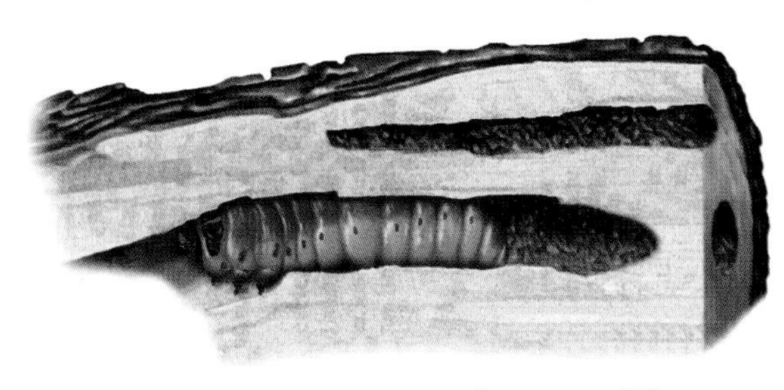

goat-moth caterpillar

common piddock

A patient mollusk

At birth, the tiny larva of the piddock attaches itself to a limestone rock. Its shell emerges gradually. The shell has small, pointed, sawtoothlike structures on one end. Patiently, the mollusk pivots until it has bored a deep hole in the rock, where it can hide from predators.

black woodpecker

A homebody

Some woodpeckers peck their homes in earth or cactuses. Others hollow out nests in tree trunks. These homes are often big enough for squirrels and owls. Black woodpeckers may live in the same home for four to six years, unless it is stolen by another animal.

Talented architects

Some unusually large homes are built by animals that are no bigger than most of their fellow creatures. The key to their success is patience. Some animals spend several months of every year building a shelter. That is not their only secret, however. Animals sometimes band together to build a home. The dwellings of these animals may last for a long time, often for several years.

A majestic nest

The golden eagle builds its huge nest at the top of a tree or on the face of a cliff. Built with branches about 6.5 feet (2 m) long, this structure can be up to 15 feet (4.5 m) wide! The energy the eagle uses to build this huge shelter is not wasted. The golden eagle, which can live for as long as 46 years, occupies the same nest for its entire life.

golden eagle

Are you curious?

There are more than thirty species of eagles. They inhabit all parts of the world except New Zealand and Antarctica. The most common species is the golden eagle, with a wingspan of 8 feet (2.5 m). It is found in Europe, Asia, North America, and Africa.

Strength in numbers

Many ants build anthills in the form of mounds. The hill of the wood ant is the most impressive. Built on an old stump, it is made of earth, conifer needles, and twigs. Under the 6.5-foot- (2 m-) high mound are several tunnels. The hill can provide shelter for up to a million ants!

wood ant

social weaver

A collective nest

Social weavers live in colonies of about 100 couples. Working in teams, they build the world's largest known collective nests. Under a large common roof, each family has its own dwelling. Made from grass and branches, the nests can be over 16 feet (5 m) in diameter. Some of these nests have been lived in for more than 100 years.

A major project

To build its nest, the mallee fowl works for 11 months a year, from morning to night. Using its feet, it digs a hole and fills it with damp leaves and branches, which it covers with a mound of sand. As it rots, the plant material releases the heat needed to incubate the eggs.

mallee fowl

More architects

Animals have amazed and inspired humans for thousands of years. Animals were the first engineers and architects. In animal societies, nothing is left to chance. Obeying laws of nature and mathematics, many animals have created wonderful natural masterpieces.

Large homes

Living in highly organized colonies, certain termites build very large dwellings called termitariums. These homes have thick walls that are made from a mixture of excrement, saliva, and earth or wood. When baked in the sun, the walls become as hard as cement. This structure includes egg compartments and food-storage areas. Fresh air flows into the dwelling through a network of chimneys.

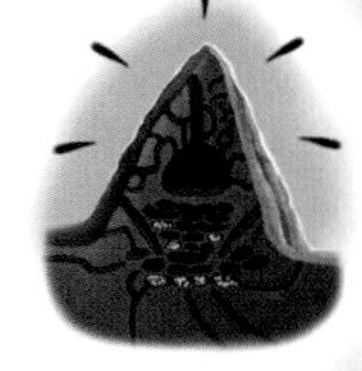

Are you curious?

People sometimes salvage the earth from abandoned termitariums to make bricks for building houses. A single termitarium in Africa is said to have provided 4,500 of these bricks!

termite

paper wasp

Cardboard houses

Wasps have been making paper for a long time. They use this material to build cardboard dwellings that can shelter up to two hundred wasps. To make their building material, wasps pull small pieces of wood from trees, houses, or fences. They mix the material with their saliva by chewing. This creates a fine pulp that dries into stiff paper.

honeybee

Wax rooms

Worker bees often build their wax dwellings in a hollow tree. Wax secreted from three tiny holes below their abdomen is used to build a hive. Each small chamber, or cell, of the hive is a hexagon. The cells are used to store the honey and the pollen or to receive eggs.

A champion builder

The beaver uses its powerful jaws and sharp teeth to chop down and strip tree trunks and branches to build its dam and lodge. The dam keeps the water level constant and allows the beaver to swim, find food, and build its lodge throughout the year. Made from branches, stones, and mud, the lodge has underwater entrances. Just above the water is a room for the family.

Canadian beaver

Temporary homes

Some animals need shelter for only a few hours a day – during sleep, for example. These animals often hide in special places. Others have a life cycle with different stages. Butterflies must first survive as eggs, larvae, and chrysalises. Some of these life-forms are very fragile! During their changes, many insects make a cocoon in which they take shelter from their enemies.

The comfort of silk

The caterpillar turns into a moth during several stages of change. Several species of caterpillars make a cocoon of silk in which they go through this transformation. The silk is produced by two special glands located at the level of their lower lip. It is secreted as a liquid and becomes solid in the air.

silkworm moth

Are you curious?

In Asia, silkworms are raised by the millions. Each silkworm cocoon contains hundreds of feet (meters) of silk thread. These threads are processed and used to make magnificent fabrics.

A cover for the night

Many parrot fish found in tropical seas wrap themselves in a cozy cover for the night. This cover is made from mucus secreted by their skin glands. It is open at the front and back to allow water to flow through it freely. This allows the fish to breathe easily. It takes the parrot fish 30 minutes to make its shelter and another 30 to dispose of it.

parrot fish

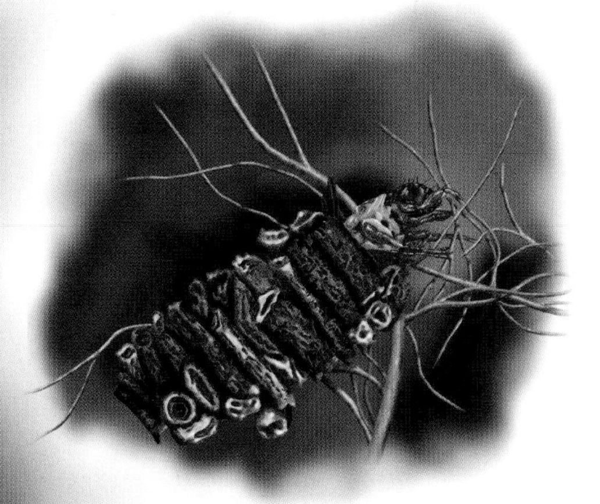

caddis fly larva

A brief existence

As soon as it is born, the caddis fly larva builds its tubelike shelter from tips of plants, fragments of shells, pieces of seaweed, and sand. Underwater, it captures and cuts up its building materials. It then glues them together with the silk it secretes. Inside its tube, it awaits its transformation. Adult caddis flies live for only the few hours it takes them to mate.

A house in the trees

Some apes make beds where they spend the night with their family. Every night, the female gorilla builds a platform of interlaced branches on which she places a small bed of leaves — just large enough for her and her child. The male builds his own bed, but he soon outgrows his treetop shelter and has to abandon it to sleep on the ground.

gorilla

No shelter at all

Not all animals are skilled builders. Many do not live in an environment that provides them with the right building materials. Some have no choice but to spend their nights under the stars and their days in the full light of the sun.

A makeshift nest

Emperor penguins live on frozen ice in Antarctica. After laying their eggs, the females return to the sea. The males must protect the eggs for the entire winter. Huddled together, the males keep the eggs from touching the ice by keeping them on top of their feet, which serve as a sort of nest.

common murre

Are you curious?

Murres live on rocky coasts in the northern hemisphere. They resemble penguins, but these two birds are not related in any way. Unlike penguins, whose wings are like large fins, murres can fly.

sea otter

Dozing on a seaweed bed

The sea otter is a marine mammal that lives in cold seas. It needs air to survive and cannot build a shelter in the water. How does it get any rest? When night falls, the sea otter lies on its back and wraps its body in seaweed, which supports it. There, not far from shore, it can rest undisturbed.

red deer

A safe place to rest

Deer have no permanent home. To rest, they choose a location where they are shielded from danger. Deer often gather in a clearing during the day. They feel safe in a group. If the location is peaceful, they will return to it. The scent of the animals becomes so strong in these clearings that human beings can detect them.

giant panda

No need for shelter

The giant panda, which lives in bamboo forests, does not have to build a shelter. Its thick, oily fur protects it from the damp and cold of the Chinese and Tibetan forests. To rest, the panda leans against a rock or a tree or rolls itself into a ball on the ground. It sleeps with its head on a carpet of conifer needles.

A map of where they live

1. **Swiftlet (p. 2)**
   (Asia)
2. **Case-bearing leaf beetle (p. 3)**
   (France)
3. **Water spider (p. 3)**
   (Europe)
4. **Lesser noddy (p. 3)**
   (Tropical islands of Atlantic, Pacific, and Indian oceans)
5. **Elf owl (p. 4)**
   (United States, Central America)
6. **Red-beaked hornbill (p. 5)**
   (Africa)
7. **Bahamian amazon (p. 5)**
   (Bahama Islands)
8. **Red fox (p. 5)**
   (Europe, Asia, North America)
9. **Trap-door spider (p. 6)**
   (South America, Central America, Mexico, the United States, Africa, Australia, Malaysia, Southeast Asia)
10. **Ghost crab (p. 7)**
    (the United States, Central America, South America)
11. **Atlantic puffin (p. 7)**
    (North Atlantic Ocean)
12. **Lugworm (p. 7)**
    (coastal areas of Europe)
13. **Rufous hornero (p. 9)**
    (South America)
14. **Tree frog (p. 9)**
    (South America)
15. **Greater flamingo (p. 9)**
    (shorelines of the Caribbean Ocean)
16. **Potter wasp (p. 9)**
    (Europe, Asia, Africa, North America)
17. **Black-tailed prairie dog (p. 10)**
    (the United States, Mexico)
18. **Eastern pocket gopher (p. 11)**
    (North America)
19. **Old World rabbit (p. 11)**
    (Europe, Africa, Australia, New Zealand , Chile)
20. **Aardvark (p. 11)**
    (Africa)

21. European mole (p. 12)
(Europe, Asia)

22. American badger (p. 13)
(Canada, the United States, Mexico)

23. Common wombat (p. 13)
(Australia, Tasmania)

24. Naked mole rat (p. 13)
(Africa)

25. Masked weaver (p. 14)
(Africa)

26. Old world harvest mouse (p. 15)
(Europe, western Russia)

27. Long-tailed tailor-bird (p. 15)
(India, Asia)

28. Weaver ant (p. 15)
(Africa, Asia)

29. Carmine bee-eater (p. 16)
(Africa)

30. Goat-moth caterpillar (p. 17)
(Europe)

31. Common piddock (p. 17)
(coasts of the English Channel, the Atlantic Ocean, the Mediterranean Sea, the Black Sea)

32. Black woodpecker (p. 17)
(Europe, Asia)

33. Golden eagle (p. 18)
(Europe, North America, Africa)

34. Wood ant (p. 19)
(Europe, North Africa, Siberia, North America)

35. Social weaver (p. 19)
(Africa)

36. Mallee fowl (p. 19)
(Australia)

37. Termite (p. 20)
(North America, Asia, Europe, Australia, Africa, South America)

38. Paper wasp (p. 21)
(Mexico, Argentina, Guinea, Trinidad)

39. Honeybee (p. 21)
(native of Asia, domesticated throughout the world)

40. Canadian beaver (p. 21)
(Alaska, Canada, the United States)

41. Silkworm (p. 22)
(native to China; cultivated in India, Japan, Spain, France, Italy)

42. Parrot fish (p. 23)
(all tropical seas)

43. Caddis fly (p. 23)
(Europe)

44. Gorilla (p. 23)
(Africa)

45. Common murre (p. 24)
(northern coasts of the Atlantic and Pacific oceans; North America, northern Europe, Greenland, Iceland)

46. Sea otter (p. 25)
(Pacific coastal waters of North America)

47. Red deer (p. 25)
(Europe, Asia)

48. Giant panda (p. 25)
(China)

More fun facts

BUILDING MATERIALS		
Group	**Animal**	**Materials**
Insects	Paper wasp	Wood fibers and saliva
	Goat-moth caterpillar	Small pieces of wood, twigs, sand, gravel, snail shells
	Weaver ants	Fresh leaves and silk
Fish	Threespine stickleback	Plant debris and sticky secretions
Birds	Woodcock	Dead leaves
	Bowerbirds	Twigs, moss, snail shells, berries, insects, flowers, leaves, mushrooms, pieces of charcoal
	Swiftlet	Saliva
	Rufous hornero	Earth
	Calliope hummingbird	Spider's web, lichen, moss, plants
Amphibians and reptiles	African tree frog	Foam produced from a thick liquid secreted by the frog
	Morelet's crocodile	Mud, branches, and rotting leaves
Mammals	Beaver	Wood, mud, stones
	Muskrat	Reeds and mud
	Squirrel	Dead leaves, twigs, moss, sod bark, lichen, feathers, wool
	Opossum	Leaves and grasses
	Gorilla	Leaves

NEST-BUILDING BIRDS AND THEIR RECORDS		
Bird	**Dimensions of nest**	**Record**
Calliope hummingbird	Diameter: 0.8 inches (2 cm) Height: 1.2 inches (3 cm)	The smallest nest
Ostrich	Diameter: 10 feet (3 m)	The nest that contains the biggest eggs
Golden eagle	Height: 6.6 feet (2 m) Diameter: 10 feet (3 m)	The largest elevated nest
Social weaver	Diameter: 16 feet (5 m)	The largest collective nest
Common scrubfowl	Diameter: 40 feet (12 m) Height: 16 feet (5 m)	The largest nest on the ground

A HOME OF ONE'S OWN		
	Animal	**Dwelling**
Wild animals	Wasp	Wasp nest
	Termite	Termitarium
	Ant	Anthill
	Bee	Hive
	Mole	Burrow
	Bird	Nest
	Rabbit	Burrow
	Hare	Form
	Fox	Den
	Beaver	Lodge
	Boar	Wallow
	Snake	Lair
	Tiger	Den
	Wolf	Den
	Bear	Den
	Bat	Cave
	Mouse	Hole
Farm animals	Horse	Stable
	Cow	Shed
	Hen	Henhouse
	Pig	Hog house
	Sheep	Sheepfold
	Dog	Kennel

For your information

swiftlet

class	Aves
order	Apodiformes
family	Apodidae
distribution	Southeast Asia
habitat	limestone caves of tropical forests
diet	spiders, insects
reproduction	1 or 2 eggs per clutch
predators	birds of prey, human beings who collect the nests

elf owl

class	Aves
order	Strigiformes
family	Strigidae
distribution	southwestern United States, Central America
habitat	arid regions with giant cacti
diet	grasshoppers, beetles, moths, caterpillars
reproduction	3 to 5 eggs per clutch
predators	no specific predators

trap-door spider

class	Arachnida
order	Araneida
family	Thera-phosidae
size	1 to 4 inches (2.5 to 11 cm)
distribution	United States, Mexico, Africa, Australia, Malaysia, Southeast Asia
habitat	dry earth
diet	insects, lizards
reproduction	500 to 1,000 eggs per summer
predators	human beings, wasps, mammals, birds, amphibians
life span	up to 25 years

rufous hornero

class	Aves
order	Passeri-formes
family	Furnariidae
size	about 7.5 inches (19 cm)
weight	2.6 ounces (75 g)
distribution	South America
habitat	wooded prairies
diet	soil insects, spiders, worms, mollusks
reproduction	3 or 4 eggs per clutch
predators	buzzards

black-tailed prairie dog

class	Mammalia
order	Rodentia
family	Sciuridae
size	15 inches (38 cm) including the tail
weight	1.5 to 3 pounds (0.7 to 1.4 kg)
distribution	United States, Mexico
habitat	dry western prairies
diet	mainly vegetarian
reproduction	2 to 10 young per litter
predators	birds of prey
life span	up to 9 years

European mole

class	Mammalia
order	Insectivora
family	Talpidae
size	5 to 8 inches (13 to 20 cm) including the tail
weight	2 to 4 ounces (60 to 120 g)
distribution	Europe, Asia
habitat	ground of woods and fields
diet	arthropods, worms
reproduction	3 to 4 young per litter
predators	owls, herons, weasels, ermines, badgers, foxes, cats
life span	3 years

masked weaver

class	Aves
order	Passeri-formes
family	Ploceidae
size	5.5 inches (14 cm)
distribution	Africa
habitat	semiarid and wooded savannahs
diet	grasses, insects
reproduction	2 or 3 eggs, sometimes 2 clutches

carmine bee-eater

class	Aves
order	Coracii-formes
family	Meropidae
size	11 inches (28 cm) including the tail
distribution	Africa
habitat	prairies with scattered trees
diet	grasshoppers, ants, wasps, other insects
reproduction	3 to 5 eggs per clutch
predators	birds of prey, lizards, snakes, man
life span	up to 7 years

golden eagle

class	Aves
order	Falconi-formes
family	Accipitridae
size	30 to 35 inches (76 to 89 cm)
weight	6 to 14.5 pounds (2.8 to 6.6 kg)
distribution	Europe, Asia, North America
habitat	mountains, cliffs, forests
diet	small mammals, birds, reptiles, carrion
reproduction	1 to 2 eggs per year
life span	25 years in the wild

termite

class	Insecta
order	Isoptera
family	there are 7 families of termites
size	0.6 to 0.9 inches (15 to 22 mm)
distribution	North America, Asia, Europe, Australia, Africa, South America
habitat	forests
diet	plant debris, wood, mushrooms
predators	aardvarks, pangolins, anteaters, armadillos, birds, aardwolves
reproduction	1 egg every 2 seconds, for 15 years or more

silkworm

class	Insecta
order	Lepidoptera
family	Bombycidae
size	2-inch (5-cm) wingspan
distribution	China, India, Japan, Spain, France, Italy
habitat	artificial breeding environment
diet	blackberry leaves for the caterpillars; the adults do not feed
reproduction	300 to 400 eggs in breeding conditions
predators	insectivores
life span	60 days

common murre

class	Aves
order	Charadrii-formes
family	Alcidae
size	16 inches (40 cm) long
weight	2.2 pounds (1 kg)
distribution	northern Atlantic and Pacific coasts of North America, Europe, Greenland, Iceland
habitat	coastal waters; nest in cliffs
diet	fish, crustaceans, mollusks, worms
reproduction	just 1 egg
predators	no specific predators

Glossary

amphibian: An animal, such as the frog, that can live on land or in water

beetle: An insect whose rear wings are protected when at rest by a second, harder pair

borer: Any of various insects, insect larvae, mollusks, or crustaceans that bore into hard material

branched: Divided into several branches or sections

burrower: Any animal that digs or tunnels through the ground

camouflage: A disguise meant to hide an animal from its enemies

chrysalis: The intermediate stage in the life cycle of some insects, between the caterpillar and the moth

colony: A group of animals that live together as part of a community

crustacean: Any of a class of animals that have shells and many pairs of legs

digger: A worker that digs, turns over, or moves earth

excrement: Waste matter discharged by the body

filament: A structure shaped like a long, thin thread

gland: An organ that produces a secretion

globe: Any sphere-shaped covering

hexagonal: Having six angles and six sides

incubation: The process of sitting on eggs to warm and hatch them

larva: An often wormlike form that is one stage of an animal's life

limestone: Rock made up mainly of calcium carbonate

mammal: A member of any animal species in which the female has mammary glands for feeding her young

marsupial: An animal whose young spend several months after birth in their mother's pouch, where they nurse from her mammary glands

mollusk: An animal with a soft body that has no bones but usually has a hard shell

mortar: A mixture of various substances builders use to bond or cover materials

mucus: A thick, transparent liquid

offspring: All of the children or young of a human or animal

predator: An animal that destroys or eats another

prey: An animal that is the victim of a predator

pulp: A soggy, pastelike substance

rodent: A mammal with sharp incisors that eats by gnawing, such as the mouse

saliva: A secretion in the mouth produced by salivary glands

savannah: A treeless plain

secrete: To form and give off from the body

secretion: A liquid substance produced by the cells of a living organism

slope: A side or an incline of a mountain

snout: The muzzle of a boar or pig

stationary: Immobile, not moving

structure: Something that is built, or constructed

termitarium: A termite's nest

tropical: Relating to or occurring in a frost-free climate with warm temperatures

twist: Something turned back or rolled back on itself

vegetation: The plant life of a particular region

wingspan: The distance between the tips of fully spread wings

Index

Editorial Director Caroline Fortin **Research and Editing** Martine Podesto **Documentation** Anne-Marie Brault, Anne-Marie Labrecque **Page Setup** Lucie Mc Brearty **Illustrations** François Escalmel, Jocelyn Gardner **Translator** Gordon Martin **Copy Editing** Veronica Schami **Gareth Stevens editing** Joan Downing **Cover Design** Joel Bucaro, Scott Krall